Gardens & Seeds

Early Childhood Theme Teaching Collection

Written by Michelle Lee Majors-Williams

Illustrated by Shelly S. Rasche

Teaching & Learning Company

1204 Buchanan St., P.O. Box 10

Carthage, IL 62321-0010

This book belongs to

____

This book is dedicated to all of those people, who throughout my life, have helped me to grow!

Special thanks goes to my husband Sam, Jeri Carroll and to my family for all of their support, encouragement and patience.

Many activities in this book require children to use seeds or beans. It is important to supervise young children during the activity to make sure the small items are used properly and do not pose a choking or other hazard.

Cover art by Shelly S. Rasche

Cover photo © Corel Corporation

ISBN No. 1-57310-105-2

Printing No. 987654321

Teaching & Learning Company
1204 Buchanan St., P.O. Box 10
Carthage, IL 62321-0010

Table of Contents

Background Information 5
10 Important Garden Facts 8
Words to Know 9
Gardens & Seeds Clip Art 10
Parent Newsletter (English) 12
Parent Newsletter (Spanish) 13
ABC Board with Pictures–Suggested
Activities 22
ABC Cards 15
ABC Card Definitions and
Descriptions 22

Math
Just How Many in a Jar? 24
A Pea for a Penny 26
More Math Activities 27

Language Arts
The Carrot Seed 28
There's a Rabbit in
My Garden 29
More Language Arts
Activities 31

Music
We're Growing in the Garden . . .32
Everything Grows 33
More Music Activities 34

Movement
We Are Growing Like Seeds! 35
Scoop the Seeds: A Relay 36
More Movement Activities 37

Self-Concept
Make a Seed Mosaic 38
Our Magic Beans 39
More Self-Concept Activities 40

Social
The Tale of Peter Rabbit 41
Growing Things Together 42
More Activities to Encourage
Working Together 44

Spatial
Put the Pumpkin on the Vine 45
Garden and Seed Puzzles 47
More Spatial Activities 48

Natural Science
Seed Planting 49
The Seeds Inside 50
More Natural Science
Activities 51

Nutrition
What Goes in Vegetable
Soup? 52
Fruit and Vegetable
Alphabet 53
More Nutrition Activities 54

Student Reproducibles
Assessment
To the Teacher 55
What I Already Know 56
What I Want to Learn 57
What I Have Learned 58
I Know! I Know!
How Gardens Grow! 59
Graph Grid 60
Garden Math Worksheet 61
Flower Worksheet 62

Suggested Literature for Children 63

Dear Teacher or Parent,

Every day, as we look around, we see things growing . . . grass, trees, flowers–and in some areas–fruits and vegetables. Learning about and watching a variety of plants grow is exciting for young children because they can see daily changes taking place. A unit about seeds and gardens easily captivates the interest and attention of children because it is relevant to them.

This unit focuses on activities involving fruit, vegetable and flower plants–the gardens they grow in and the seeds they grow from. It's a colorful unit using the senses as children get to see, smell, taste and touch a variety of reds, yellows, greens and purples in fruits, vegetables and flowers. It's an exciting unit as children watch seeds change and grow properly. Children will learn that responsibility and cooperation go hand-in-hand to produce successful crops and gardens.

The unit provides teachers with a wide range of developmentally appropriate activities, and it provides children with exciting hands-on learning opportunities. Activities incorporate multiple subject areas and are integrated throughout the unit.

Many activities can easily be used at centers in your classroom. Create exciting centers suited to your children where they can work independently or in small groups. You may want to prepare task cards that clearly list the steps, directions or instructions for children to follow, and for the youngest learners illustrated task cards will be helpful.

Familiar children's literature is suggested and used throughout the unit either as the focus of an activity or to enrich and extend it. Add your own favorite titles as they relate to specific topics and lessons.

Add excitement to this unit and extend beyond the classroom! Enlist the expertise of parents and local merchants who have "green thumbs" and whose businesses focus on growing plants, flowers, fruits and vegetables. Arrange visits to a local nursery, greenhouse or florist to learn about flowers from a professional. Let children see the ideal conditions plants need in order to bloom. Have them smell flowers as they are growing. Visit a local produce market or the produce section of your supermarket and see which fruits and vegetables children are familiar with. Introduce new ones! If possible, visit a vegetable farm or orchard and see fruits and vegetables growing and being picked when they are ready to go to market.

Make children aware of special gardens in your community. Visit the formal gardens in season at historical sights, museums or municipal buildings in your city or town and see special garden displays at your local parks.

"Dig" into this unit and watch the learning experiences and your students grow and grow!

Sincerely,

Michelle

Michelle Lee Majors-Williams

Background Information

Seeds

More than half of all kinds of plants in the world grow from seeds. Some seeds, such as coconuts, are big while some seeds, such as mustard seeds, are tiny. Seeds may be different in color, shape and size. However, all seeds are alike in *three* ways.

- They all have a baby plant inside called an *embryo.*
- All seeds have some food for the baby plant called *cotyledon.* This food helps the baby plant until it can function independently.
- All seeds contain a seed coat on the outside which protects the baby plant and its food.

Almost all trees, shrubs, flowers and vegetables start as seeds. Some seeds grow into *seedlings* and then into adult plants. Plants produce an enormous number of seeds to insure reproduction of the plant.

Seeds travel in several ways. Some seeds travel by land, some by water and some by air. Many of the seeds that travel by air have *parachutes.* Others have wings. Most of the seeds that travel by water are light enough to float. Some seeds that travel by land are *hitchhikers.* They may have sharp prickles that help them catch onto the fur of animals. Some seeds may get *rides* in mud that stick to the feet of animals and birds. Birds often carry seeds hundreds of miles in the mud on their feet. And some seeds get themselves scattered by being in fruits that birds like. The birds eat the soft part of the fruit and drop the seeds.

Tumbleweeds have a different way of scattering their seeds. The whole plant dries up and rolls along the ground dropping its seeds as it goes along. *Witch hazel, touch-me-not* and some other plants shoot their seeds out of their seed pods. The best "shooter" is the South American plant called the *monkey's dinner bell.*

We eat a variety of seeds. The food stored in seeds for the baby plant makes nutritious food for us. Among the seeds we eat are beans, peas and nuts. Wheat, oats, corn and rice are different kinds of grasses called *cereals.* The seeds of these cereals provide food for many animals–some of which provide food for people such as cows and pigs.

When seeds are ripe, they drop from a parent plant. Given water, air and warmth, they will begin to grow. A root will appear first and begin to grow downward. Then a stem pushes upward towards the light and a plant is on its way to being productive.

Gardens

Gardening is the number one leisure time activity in the United States and is very popular in Canada as well. There are three main types of gardens: vegetable, flower and indoor plant gardens.

Vegetable Gardens

Vegetables that are common in today's gardens originally came from different parts of the world. Tomatoes, potatoes, peppers, squash, string beans and lima beans came from South America. Celery, radishes, lettuce, peas, spinach, asparagus and beets are from countries near the Mediterranean Sea. Carrots, cabbage, parsnips and turnips originated from Europe, while artichokes, okra and yams are from Africa. Almost none of the vegetables that are familiar to us today are native to North America.

A vegetable is the edible part of a plant. It may be the root, bulb, stem, leaves, seeds, flower, or it may be the *fruit*–the part of the plant that contains the seeds.

Gardens may contain vegetables that grow *above ground* and *below ground.* Tomatoes, beans, lettuce and peas grow above ground. Carrots, onions, radishes and potatoes grow below ground.

Flower Gardens

Some of the flowers that are grown from seeds are zinnias, nasturtiums, petunias and cornflowers. These flowers are annuals–they have to be planted every year. Other flowers such as black-eyed Susans, Shasta daisies and phlox are *perennials*–once planted they will bloom in season for at least three years without replanting. And some flowers, like hollyhocks, are *biennials*; meaning they bloom the year after they are planted.

Some garden flowers are very particular about the type of soil they grow in. Sweet peas will not grow well unless they have a very rich soil. Zinnias can grow well in poor soil but require sunshine and will not grow well in shaded areas.

Popular favorites in many springtime gardens are tulips and daffodils which grow from bulbs. These originated in Turkey and are now produced mostly in the Netherlands.

It is important to note that growing conditions vary greatly depending on the region of the country. Many plant and bulb catalogs define growing regions by *zones* and will recommend specific flowers that will survive in each zone.

Today's flower gardens have a variety of flowers that originated from many countries around the world.

10 Important Garden Facts

1. A garden is a place where flowers, fruits and vegetables are grown.
2. Gardens come in many different shapes and sizes.
3. Plants in a garden need to be tended and cared for.
4. Plants need water, air, sun and space to grow.
5. Many new plants grow from seeds.
6. A fruit is the part of a plant that contains seeds.
7. Many fruits are good to eat.
8. A vegetable is the part of a plant that can be eaten (stem, leaf, root, etc.).
9. Insects and animals can live in gardens.
10. Some insects and animals are helpful to the plants in a garden. Some are not.

To the teacher: These basic gardening facts may be enlarged and made into a class chart and reinforced as the unit progresses.

Words to Know

Gardens
care
flower
indoor gardens
insects
outdoor gardens
planting
rows
soil
vegetable
weeds

Above Ground
beans
flower
leaf
lettuce
peas
stem
tomatoes

Below Ground
bulb
carrots
onions
potatoes
radishes
root

Tools
hoe
hose
rake
rototiller
shovel

Seeds
edible seeds
grow
kinds
matching
parts
plant
sorting

Seed Travel
air
hitchhikers
land
water

Plant Growth
air
embryo
leaves
seed
seedling
space
stem
sun
water

Plant Uses
decoration
food
homes
shade

To the teacher: Use these groups of words to increase your students' working vocabulary about gardens and seeds. Create an illustrated chart or have students make individual picture dictionaries.

Gardens & Seeds Clip Art

Use clip art to create charts, graphs, student worksheets, class or individual journals and for bulletin board displays.

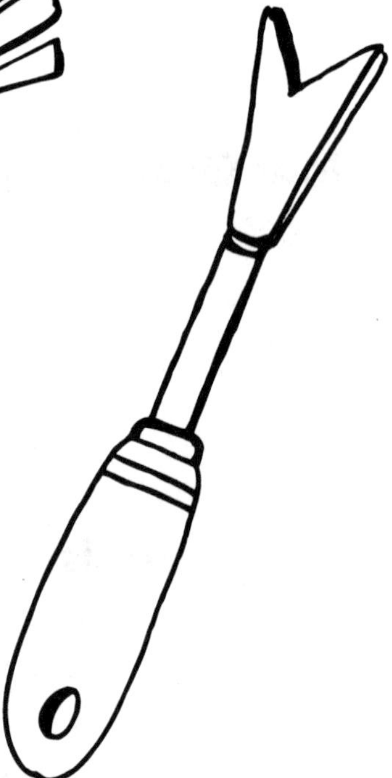

Gardens & Seeds

Dear Parents,

Today was the first day of our unit about gardens and seeds. We are all excited about the fun we are going to have. Each day, please ask your child what he or she did at school. Some of the activities we will be doing include:

sampling fruits and vegetables and making a variety of graphs

sorting seeds by color, shape and size

learning about gardens and how to take care of them

planting our own seeds and watching them grow

Each child will be keeping a seed log and writing and drawing about what is happening to his or her seed.

At home, please discuss some of the growth that may be taking place in your own yard and garden. Or plan a family outing and include a visit to a local garden or park to see what is in season. Take a nature walk and see what is growing in and around your neighborhood. Thank you for your support.

Sincerely,

Please provide information regarding any food allergies or restrictions as we may be tasting different seeds, nuts, fruits and vegetables.

Las Noticias de Nuestra Sala de Clase

Jardines Semillas

Estimados Padres,

Hoy ha sido el primer día de estudiar las semillas y los jardínes. Todos estamos muy illusionados por lo bien que lo vamos a pasar. Agradeceríamos si le preguntaran a su hijo/a sobre lo que hizo él o ella en la clase ese día. Algunas de las actividades que hemos planeados incluyen.

probar una variedad de frutas y verduras y hacer grácficos para reflejar nuestras preferencias

separar las semillas por su color, forma, y tamaño

aprender acerca de los jardínes y como cuidarlos

sembrar nuestras propias semillas y observarlas crecer

Cada alumno va a mantener un diario de apuntes escribiendo y dibujando los cambios que observa en las semillas.

En la casa, discute por favor algún crecimiento que puede verificarse en tu yarda y en tu jardin. O planea una excursión con tu familia y incluye una visita a un jardín o parque local para ver que está en estacion. Da un paseo natural y vea que crece en y alrededor de tu barrio. Garcias por tu sostén.

Sinceramente,

Haga el favor de dar la información respecto a alergias o restricciones a cualquiera comida porque es posible que podamos probar una variedad de semillas, nueces frutas y vegetales.

ABC Board with Pictures

For this activity, 26 alphabet cards relating to gardens and seeds are provided. The cards may be used in a variety of ways. Adapt them depending on the age and developmental level of your children.

Materials

poster board
26 library pockets
markers to write letters or
 ready-made letters

Suggested Activities

- Use the cards to reinforce alphabet letters and letter sounds.
- Duplicate the cards to create a Garden and Seed Memory Match Game.
- With a group of four to six children, have each child choose his or her favorite card. Talk about why he or she chose that card.
- Children can use the cards as a spelling aid in their writing.
- Hold up any given letter card. Ask students to name another *garden* and *seed* word beginning with the same letter.
- Hold up two or three letter cards in sequence–j, k, l. Ask students to name the letter that comes next.
- Mix up the cards and have students rearrange them in alphabetical order.
- Hold up a card and ask students to name a rhyming word.

Aa

Bb

Cc

Dd

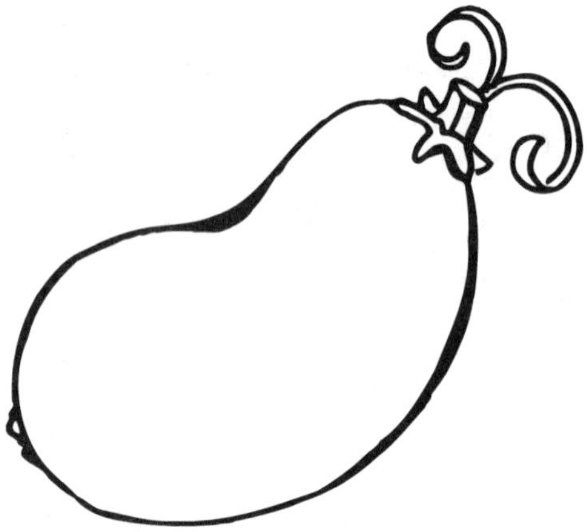

Ee

Ff

Gg

Hh

Ii

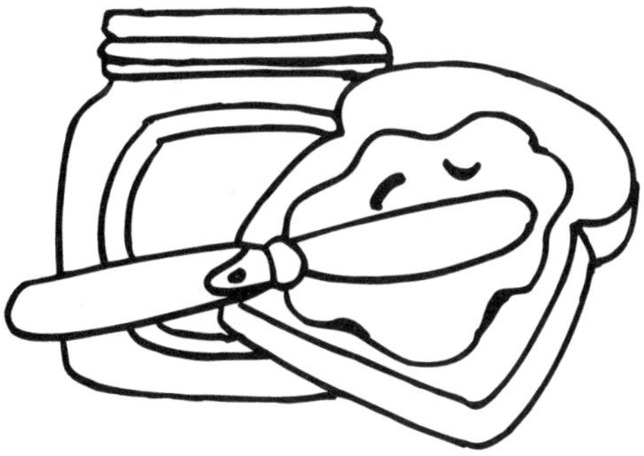

Jj

Kk

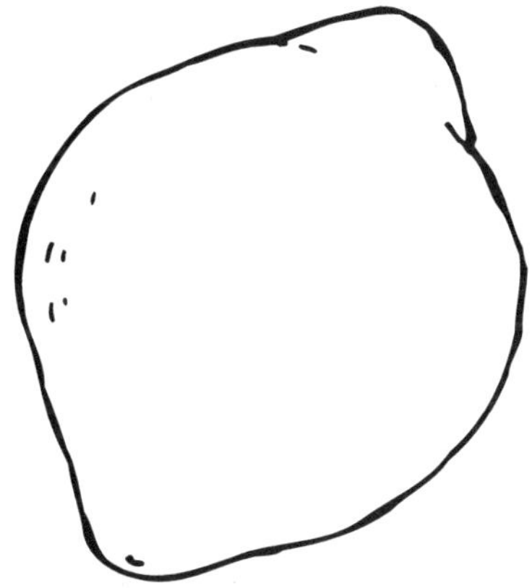

Ll

Mm

Nn

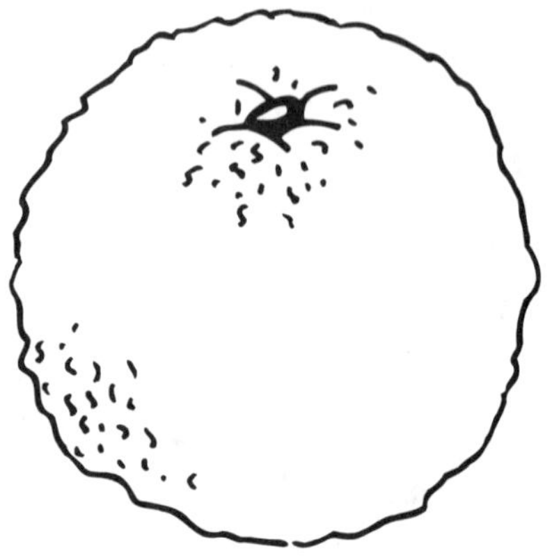

Oo

Pp

Qq

Rr

Ss

Tt

Uu

Vv

Ww

Xx

Yy

Zz

ch

gr

ABC Card Definitions and Descriptions

An **apple** is a fruit that grows on trees. It can be red, yellow or green.

A **garden** is a piece of ground where flowers and vegetables are grown.

A **banana** is a yellow curved fruit that grows in bunches on tall trees.

A **hoe** is a garden tool used to break up the soil.

A **carrot** is a long, orange vegetable. It grows under the ground.

Indian corn is a type of corn that has many different colors of kernels.

Dirt is earth that can be used to plant seeds in.

Jam is a mixture of fruit pulp boiled with sugar.

An **eggplant** is an oval-shaped vegetable with a deep purple skin and tan color inside. It grows on a vine.

Kale is a member of the cabbage family and has loose green leaves.

A **flower** is the part of a plant with brightly colored petals. They come in all shapes, colors and sizes.

A **lemon** is a yellow citrus fruit that grows on trees.

To the teacher: Cut out each word description and paste on the back of the corresponding letter card.

Melons like muskmelon, watermelon and cantaloupe grow on vines.

An **ugli** is a fruit that has a very hard bumpy outside but has a good-tasting inside.

A **nut** is a seed with a hard shell outside and a kernel inside. Nuts usually grow on trees or bushes.

A **vegetable** is a plant used for food. Vegetables are usually grown in gardens.

An **orange** is a citrus fruit that grows on trees.

Weeds are plants that seem to grow anywhere but are not wanted in gardens. They crowd out the vegetables and fruits.

A **pinecone** carries seeds which will grow into pine trees.

Xigua (pronounced she-gwah) is the name for *watermelon* in Chinese. The seeds can be dried and the center can be eaten.

A **quince** is a golden or greenish-yellow apple-shaped fruit with a hard flesh inside used to make jam.

A **yam** is a vegetable that grows underground. It is a little bit like a sweet potato.

Rain is the drops of water that fall from clouds to the Earth.

A **zucchini** is a long, thin vegetable that is green on the outside but creamy colored on the inside. It grows on a vine.

The **sun** provides heat and light for plants to grow and to keep us warm, too.

Cherries are small fruits and have pits in their centers.

A **tomato** is a round red fruit that has seeds. It grows on a vine.

Green grapes and purple grapes are often the favorites of children. They can taste sweet or sour. Some have seeds and some do not.

Just How Many in a Jar?

Students will observe, predict, estimate and count objects in a jar.

Materials

estimation jar
estimation box
small pieces of paper to write students' guesses on
reproducible on page 25
pencils
beans to go inside estimation jar

Group Size The whole class can participate.

In Advance Gather all of the materials needed.

Activity Agenda Show students a small jar filled with lima beans or any other type of dried beans. Ask them to guess how many they think are in the jar together–aloud. Look at the estimates each student made. Which child/children came closest to the actual number of beans in the jar? Did anyone guess the exact number?

- Repeat the activity using different sized jars.
- Repeat the activity using smaller and larger beans.
- Bring in jars of items such as olives, pickles and maraschino cherries. Have students estimate how many items are in the jar. After counting the actual number and seeing who came closest, have a tasting session and enjoy the estimating treat!

Closing Talk about how trees grow from seeds. Read *A Tree Is Nice* by Marc Simont.

Name ____________________

Date ____________

Estimation Jar

Today there are __________ in the jar.

My guess is there are __________ things inside.

The actual number inside the jar was __________.

Penny

Use with "A Pea for a Penny" lesson.

A Pea for a Penny

Children will manipulate small objects, count orally from 1 to 10 and observe cultural diversity through literature.

Materials

Picking Peas for a Penny by Angela Medearis
small plastic bags containing about 20 peas in each
counting cups
enlarged paper penny (page 25)
set of number cards from 1-10
paper
glue
variety of other seeds such as pumpkin, watermelon and apple seeds are suggested

Group Size

Groups of four to eight children at a time are suggested.

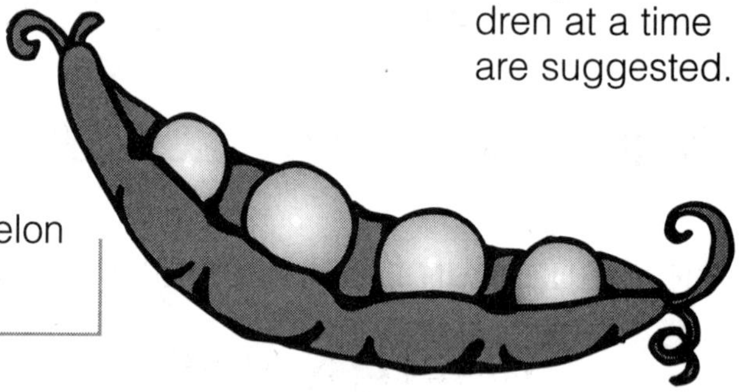

Activity Agenda

Read *Picking Peas for a Penny* to the children. Encourage children to count with you on the counting parts of the story. Show the students a real penny and the enlarged paper penny as visuals. Ask the students, "How much is a penny worth? Five pennies? Ten pennies?"

Tell the children that they will also get to pick up peas. Show the plastic bags of peas and the number cards on the table at the center. Each student will go to the table, pick up a number card and write the number on a piece of paper. Then each child counts out the same number of peas from the bag and glues them to the paper with the number.

After all children have had their turn, ask: "How many children counted three peas, eight peas, four peas, etc. How many peas were counted altogether? If you had one penny for each pea that every child counted, how many pennies would there be?"

Extended Activities

- Bring in fresh pea pods. Open them. How many peas in two pods? Ten pods? Fifteen pods? Twenty pods?
- Have the children draw pea pods with different numbers of peas in each pod?
- Have children compare tastes between fresh peas and cooked warm peas. Which do they prefer? Compare results between those who prefer fresh peas to cooked peas.
- Call out a number and call two to three children to go to the table and count out the number using the watermelon, apple and pumpkin seeds. Have each child count their seeds with the class. Repeat for numbers 1-10.

More Math Activities

Fraction Foods

Cut apples and oranges in half or in quarters. Have children find the matching half or the other three parts to make a whole. Read *Eating Fractions*.

Visit the Supermarket

Arrange to visit the produce section of a local supermarket. See how fruits and vegetables arrive. How many apples came in a carton? How many cantaloupes come in a carton? How many apples in a pound? Create your own questions and let students discover new taste treats in the produce section.

Measure, Weigh and Compare

Have a variety of fruits available. Place a scale in the classroom. Have children weigh each fruit. Let them write sentences using words and pictures.

What's the Cost?

Set up a grocery center with play foods or real fruits and vegetables, play money, grocery lists, pencils and sacks. Have the children check and record the cost of produce, select and buy items.

Fruit and Vegetable Graphs

Have children bring a variety of fruits. Create a Favorite Fruit graph. Write the name of each fruit and create an illustration as well. What is the most popular fruit?

Repeat graph activity using vegetables. What is the favorite vegetable? What vegetable(s) are least favorite? (Use graph grid on page 60.)

The Carrot Seed

Students will develop prediction and listening skills as they learn about seeds and gardens.

Materials

whole carrot
carrot coins (slices) for student sampling
large carrot cut-out
The Carrot Seed by Ruth Krauss
variety of seed packets

Activity Agenda

Gather the children around you. Hold up the carrot and ask if anyone knows how a carrot grows. Talk to the students about how some vegetables grow underneath the ground. Then show the students a variety of seed packets. Use the following questions in your discussion:

Which of these packets of seeds would you like to plant?

Where would be a good place to plant the seeds?

How would you take care of the seeds to make them grow?

How long do you think it would take these seeds to grow into plants?

Read *The Carrot Seed*. Ask the students to pay close attention to the pictures. As you read, put your hand under each word. Stop to allow children to make predictions as to what the boy's seed will turn into. Talk to the children about how the boy cared for his plant. Then discuss what seeds need in order to grow. Have children look through the book to identify as many as they can.

Extended Activities

- Write *What Seeds Need* and brainstorm the list with the children on an oversized carrot cut-out.
- Let each student who would like to have a carrot coin (slice) taste one. (Closely supervise the use of small food items.)
- Add a set of books and the tape of *The Carrot Seed* to the listening center.
- Have children look at the pictures retelling the story to themselves substituting *radish* for the word *carrot*.

Planting a Rainbow

Children will sort, count and make graphs.

Materials

Planting a Rainbow by Lois Ehlert

reproducible on page 30 for each student or each pair of students

8 assorted bags of red, blue, yellow, orange, green and purple colored plastic counting chips, or you may substitute small squares or flower cutouts of colored construction paper for the colored counting chips

crayons and markers of the same colors used for the graph of each child

picture of a rainbow showing different colors

Group Size

The whole class may listen to the story at the same time. For the graphing activity, have students work at a center of about six to eight children at a time. Children may pair up and work cooperatively on their graphs.

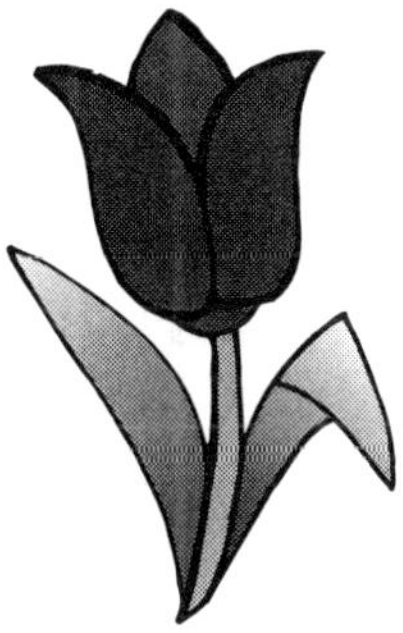

Activity Agenda

Gather the children around you and show them the picture of a rainbow. Talk with them about some of the colors that are in a rainbow. Ask the children if they have ever seen a rainbow. Show the students the book *Planting a Rainbow*. Ask the students why they think this book has the title. Read the book to the children. Tell the children that they are going to make a graph that shows different colors. Show the children the bag with the different colored chips inside. Assign the six to eight children to work in this center first. Each child will have a graph and a bag of counting chips. Children will take the colored chips out of the bag, sort them by color and then show their results using the same colors of crayons or markers. After completing the sorting activity, ask the children questions such as: How many red flowers grew in your garden on the graph? How many yellow flowers grew in your garden? Etc.

Name ______________________________

Planting a Rainbow

More Language Arts Activities

ABC Books

As a group, make a large class gardening ABC book. Or have children make individual fruit, flower, vegetable or garden theme ABC books. Have on hand a variety of magazines such as *Family Circle* and *Woman's Day* with photographs of food. Also bring in newspaper ads in color and gardening catalogs for children to cut pictures from. Supply scissors, glue, pencils, crayons and markers for writing the letter and picture names on each page. Have children exchange their books with classmates and see how they are different.

Show and Share

Have children bring their favorite fruit or vegetable to class for a tasting party. Have them show the food and share their favorite way to fix it. Make jam from fresh fruit.

Garden Riddles

After children are familiar with the vocabulary of the unit, create riddles and have them guess what you are thinking about. Give a color cue, shape cue, where they grow and what products are made out of the fruit or vegetable. "I am orange. I am round. I grow on a vine. I get picked in October." "I am yellow. I am bitter. Don't bite into me!"

Books and Tapes

Add books and tapes of stories such as *The Carrot Seed* to the listening center.

Parent Interviews

Have children interview their parents about their favorite gardening activities and share those with the group. Make arrangements to visit Jimmy's aunt's tomato garden or Elise's grandmother's flower garden.

We're Growing in the Garden

Children will increase their musical awareness while working on garden vocabulary, large motor, and pre-reading skills.

Materials

copy of the song for teacher use
poster with song words written on it
optional–plastic vegetables and children's garden hoes

Group Size

The whole group of children may participate

Activity Agenda

Have all of the children gather around you. Tell them that they are going to learn a garden song that has the same tune as a nursery rhyme they might know. Sing the song to the children and model the actions. Then practice the song all together. When the children have sung a verse several times, then you may want to point to the words on a poster as they sing and do actions.

We're Growing in the Garden

To the tune of "Here We Go 'Round the Mulberry Bush"

This is the way we hoe the soil. Hoe the soil. Hoe the soil. This is the way we hoe the soil. To get our garden ready.	(Stand up, pretend to hold a hoe, then make raking motions

Other Verses

This is the way we plant the seeds.	(Bend down and pretend to be planting seeds.)
This is the way we water the garden.	(Pretend to be holding a watering can and watering the garden.)
This is the way we pick the weeds.	(Bend down and pretend to pick weeds.)
This is the way we harvest the garden.	(Pretend to be picking up, digging up and pulling up vegetables.)
This is the way we eat the food. Eat the food. Eat the food. This is the way we eat the food. That we've grown in our garden.	(Pretend to be eating food, and take your hand and rub your stomach.)

Extended Activities

Put the song poster up in the classroom and use it when you sing the song. Encourage children to read it independently.

Everything Grows

To help students develop music and listening skills;
to help students develop word/picture recognition.

Materials

Everything Grows book and record by Raffi
record player
chart

Group Size

Whole group, then small groups could be put at the listening center

In Advance

Gather all of the materials needed.

Activity Agenda

Gather all of the children around you. Have the students think about things that grow. Ask for student responses and record them on a chart. Then read the book *Everything Grows.* After reading the book, you may want to add to your list of things that grow. Ask the students if they think we could put this book to music and sing it. Show students the song page in the back of the book. Play the song "Everything Grows." Invite the children to join in and sing. Use the book as the song is playing to help students both visually and auditorally pick up on the words and music.

This song can be used in a variety of units since it talks about many different things growing!

Closing

Ask the students to look for things that are growing on their way home from school today.

Extended Activities

- Put book and tape in listening center.
- Have students draw pictures of things that grow in a garden.
- Have students bring books from home about things that grow.

More Music Activities

Oats, Peas, Beans and Barley Grow

Sing the song "Oats, Peas, Beans and Barley Grow." Write the words to the song for use on a pocket chart.

Mulberry Bush

Sing the traditional song "Here We Go 'Round the Mulberry Bush." Make up verses to go with the tune as you plant seeds, cut fruits and vegetables, taste new foods.

This is the way we plant our seeds.
This is the way we water our seeds.
This is the way we tend our seeds.

This is the way we cut our food.
This is the way we eat our food.
This is the way we clean up our food.

Song Writers

Let the children pretend to be song writers. Encourage them to sing songs about what they are doing in their gardens, with their food or about a specific vegetable or fruit.

We Are Growing Like Seeds!

Children will use large muscle movement; children will use creative thinking and listening skills.

Materials

chart to record student responses
poster that shows the different movements throughout the poem

Group Size

The whole group of students may participate.

Activity Agenda

Have the children gather together in front of you. Tell them that they will be pretending to be seeds and to think about what kind of vegetable or flower they would like to grow into.

Invite the children to curl up **(crouch down)** on the floor and pretend that they are a little seed. Say, "All of us are little seeds underneath the ground." "Then one day the sun shines on us a little and we begin to grow." **(Students start to let one of their arms stretch upward.)** Then say, "Oh, now it is starting to rain and our roots get watered and we grow a little more." **(Students raise their other arm/hand towards the sky.)** Teacher: Then someone comes and pulls all of the weeds away from us and we grow a little more. **(Students let top half of their body stretch upward towards the sky.)** Then we get more sunshine and we grow a little, and we get rained on and we grow a little more. **(As each item is read, students slowly stretch until they are standing up, stretched out as high as they can be.)** Then say, "Oh, look at all of the wonderful things we have grown into. I grew into a ______________________." The teacher can ask each student what he or she grew into and record their answers on a chart. As each child says what he or she grew into, have him or her sit down.

The poster showing the pretend steps of growing may be displayed in the dramatic play center. Encourage students to pretend they are growing and say the steps to each other.

Scoop the Seeds: A Relay

Children will use social skills, large muscle skills, eye/hand coordination and balance in scooping beans or seeds and moving them from one container to another.

Materials

large seeds or beans to fill four same-size containers
four tablespoons
four containers filled with seeds or beans
four empty containers to transfer seeds to

Group Size Whole group, then small groups could work at the listening center

In Advance Gather all of the materials needed.

Activity Agenda The object of the relay game is to work as a team to transfer seeds in a spoon from one container to another.

Directions Divide students into four teams. At the signal, which will be when you say, "Garden," the first child standing in each of four lines scoops some seeds from a container placed on a chair next to the team. The child carefully carries the spoon and puts the seeds into a container on a chair placed about 12 feet away. The child brings the empty spoon back and gives it to the next child on their team. As each student completes his turn, he goes to the end of the line.

Repeat until each team has emptied one cup and filled the other with the seeds or until one team is first in emptying a container of seeds.

Children standing in line should cheer for their team player who is carrying the spoonful of seeds.

- Students may not run–only walking is permitted.
- Decide whether you want students to stop and pick up a seed that falls from the spoon or whether you want them to ignore it and continue on.
- Any team that has worked well together in good team spirit is a winning team.

Closing Ask the students to look for things that are growing on their way home from school today.

More Movement Activities

Potato Sack Walk

Divide children into pairs and give each pair of children a potato sack or pillowcase. Have each pair try to get from one given place to another after they have each put one leg inside the sack. Try walking then jumping.

Fruit and Vegetable Twister

Play the game of Twister® and make a Twister® mat using a white sheet. Use a hot glue gun to attach orange and red circles to depict apples and oranges. Play the game with two or three children at a time. Children place their hands on the apples and their feet on the oranges one at a time.

Potato Roll

Provide children with large potatoes. Have them roll the potatoes from one end of the room to the other using their hands, sticks, plastic shovels, etc. Which way is easiest? Hardest?

Down the Garden Path

Provide large pieces of cardboard and have small groups of children illustrate different parts of a large garden such as the carrot path, the tomato garden, the flower garden. Lay these out on the playground and leave wide strips between the pieces to form paths. Have children hop through the path; then jump through the path.

Over, Under, in and Out

Set up an obstacle course that requires children to go over some things (chairs), under some (tables), through some (tunnels) and around others (desks).

Make a Seed Mosaic

Students will work together to create seed mosaics by using a variety of seeds and beans.

Materials

assortment of seeds and beans
colored construction paper
glue
cotton swabs for spreading glue
containers for seeds
(before beginning this activity with the children, create one to two samples of a seed mosaic. Use simple patterns of fruits and vegetables that children can easily identify.)

Group Size

Small groups of four to five students may work together at a center.

Activity Agenda

Gather the whole class and show your teacher-made models. Explain that the design was made from different seeds and beans. Encourage children to be creative in the design they choose to make and not to copy your work. They may create a free-form random design, or they may want to create a definite item such as a slice of watermelon. Or they may want to write their name using dark seeds or beans against a background of light seeds.

Students may draw a picture first and then glue their seeds onto the picture, or they may wish to be spontaneous and start gluing seeds and let a design form.

Have each student describe his design; then display all seed mosaics.

Extended Activities

- Have students bring seeds and beans from home to contribute for this activity.
- Weigh each mosaic. Whose mosaic is the lightest? Whose is the heaviest?
- Count the seeds in one mosaic pattern.

Our Magic Beans

Young children will enjoy guessing what their beans will grow into.

Children will listen to this traditional folktale and use their creativity and individually to write and illustrate what their magical beans will turn into.

Materials

Jack and the Beanstalk by Matt Faulkner (or any other well-told version)
Jack and the Beanstalk bulletin board–provide a background of light blue construction paper for the sky. Draw a large green vine spreading across the background.
glue
cotton swabs for spreading glue
crayons and pencils
large bean-shaped chart to record vocabulary words on

Group Size The whole class may participate.

Activity Agenda

Read *Jack and the Beanstalk* to the children. Explain that they will get to be like Jack and also grow something from magical beans. Show students your teacher-made vine example but not what might grow on the vine.

Brainstorm with the students many ideas of what the beans could grow into accepting all creative answers. Talk about how the beans aren't magical, and that we are just pretending. That is why Jack and the Beanstalk is called a folktale. List vocabulary words on a bean-shaped chart for children to refer to in their writing activity.

Have each child glue beans onto a piece of paper and write two to three sentences about what their beans grew into. Attach each child's writing sample to the vine. Title the bulletin board *Our beans grew into . . .*

More Self-Concept Activities

Compare Garden Favorites with Friends and Family

Encourage children to ask their friends and family members what their favorite fruit, flower and vegetable is. Compare it to their own favorites. Prepare a sheet for children to write their favorites and to record the responses of friends and family.

Encourage children to return completed responses. Analyze and compare results by graphing or tallying one or more of the choices.

	Mine	**Person 1**	**Person 2**
Favorite Fruit			
Favorite Vegetable			
Favorite Flower			
Favorite Color Flower			

Servings of My Favorite Foods

Give each student a paper plate, magazines, newspapers, grocery ads, scissors and glue. Have them create servings of their favorite foods on the plates. For those who are interested, have them decorate the plates.

The Tale of Peter Rabbit

To reinforce the concept of obeying and following rules of good behavior.

Materials

The Tale of Peter Rabbit by Beatrix Potter

Group Size

The whole class may participate.

Activity Agenda

Gather children around you and read this all-time favorite children's story about Peter, Flopsy, Mopsy and Cottontail. The latter are good little bunnies, but Peter disobeys his mother and goes to Mr. MacGregor's garden with nearly disastrous consequences.

As one of the best-known and best-loved stories for children, *The Tale of Peter Rabbit* is dramatic, exciting and complete–all on a very small scale. The watercolor illustrations are exquisite; be sure to share them closely with the children.

After you read the story, ask the children to share some stories of what happened to them when they disobeyed.

Directions

- Have children reenact and dramatize particular scenes from the story.
- Set up a library center with companion titles such as: *The Tale of Benjamin Bunny, The Tale of Squirrel Nutkin* and *The Tale of Jemima Puddle-Duck* and other suitable titles that are available.
- Bring in other works of Beatrix Potter and have children study her watercolor illustrations.
- Have students work in small groups and experiment with painting with watercolors.
- Make a class chart listing how rabbits cause problems in gardens.

Growing Things Together

Depending on the season and the growing conditions of the geographic region where you are located, select one or more of the following activities. These activities focus on children working together in pairs or small groups to create a new plant. Individual effort and responsibility is important, but the combined group effort to plant, monitor and care for the seeds and bulbs should be the main concern.

Sowing Seeds Fill a seed tray with potting soil. Sprinkle the seeds over it; then cover them lightly with more soil. Water the seeds gently using a watering can with a fine spray. Tie a large clear, plastic bag over the tray and put it in a warm, dark place. When the plants have six leaves, take them out of the closet and replant 2" (5 cm) apart in new trays so they have more room to grow. When the weather is warm enough, move the trays outdoors. Once the plants have grown too big for the trays, dig them up very carefully keeping the soil around the root of each plant. Dig small holes in the garden and carefully plant the seedlings in them. Press the soil down around seedlings and water them.

Growing Sunflowers

Giant sunflowers grow very fast. You can plant the seeds outdoors in the spring. Choose a sunny place with good soil, near a fence or wall if possible. Pull all the weeds. Dig and rake the ground so the soil is level and does not have any lumps. Push the sunflower seeds into the ground about 1/2" (1.25 cm) deep and 2' (.61 m) apart so they have room to grow. Water the ground every day and shoots should appear in 10 to 14 days. The plants will grow very fast if you remember to water them. Have a group contest to see whose sunflowers grow the fastest and tallest! Once the sunflowers are growing, tie each one to a stick or stake to keep from flopping over as sunflowers grow several feet tall. Measure them every week and have each pair or group of children keep a chart of their sunflower's growth. Sunflowers bloom from June to September; therefore this is an ideal summer-time activity. When the flowers die, cut off one of the heads and remove the seeds and lay on newspaper with a fork. Put the seeds in an envelope, seal, label it and keep it in a dry place. Plant the seeds next spring. Let the children notice how sunflowers turn to face the sun!

Planting Bulbs

Bulbs are easy to grow indoors. Plant crocuses, daffodils or tulips in the fall and they will bloom in spring. Fill clay pots with potting soil. Put the bulbs on top with the pointed end up and close together. Fill the pots with more soil. Small bulbs must be covered. Big ones can poke out of the soil. Water them. Put the pots in a cool, dark place for eight to 10 weeks. Check the soil occasionally to see that it is damp. When the shoots are about 2" (5 cm) high, put the pots in a light, cool place. Water the plants regularly. When they are taller, move them to a warmer, light place. Use stakes to support any droopy plants. Once the flowers bloom and die, cut off the heads but not the leaves. Cut the leaves back only after they have completely died and have sent the food back to the bulbs. Note: Plant the bulbs outdoors next season, if possible, as they will not flower indoors a second time.

More Activities to Encourage Working Together

Jack and the Beanstalk

Start a Jack and the Beanstalk book. Write the words *Once upon a time* on the first page. Have each student write one sentence to continue the story. If needed, children may have a second turn to write until the story reaches its end. Read the story to the children and discuss how everyone worked together to write the story.

Fruit and Vegetable Puppets

Provide colored felt, wiggly eyes, pipe cleaners, markers, fringe, rickrack and fabric scraps. Have children work in small groups to create fruit and vegetable puppets. Encourage students to work together to present a garden play using their puppets.

Make a Garden Mural

Display a variety of fruits and vegetables and ask children to describe where each one grows. Encourage children to tell if a particular vegetable grows above or below ground. Provide large butcher paper affixed to a bulletin board or wall along with felt, fabric scraps, markers, crayons, glue, construction paper and Easter grass. Have small groups work together on the mural to create garden scenes depicting what they have learned about seeds and plants, fruits, flowers and vegetables. Invite another class to see the mural.

Read *Oh, Bother! Someone Won't Share* by Betty Birney. Rabbit won't share, but the animal friends come to help protect his garden anyway. Rabbit decides that sharing is a good thing to do. Have children bring a toy, doll or game from home to share with a classmate.

Pumpkin, Pumpkin

Children will learn how and where pumpkins grow and play a related version of the familiar game Pin the Tail on the Donkey.

Materials

Pumpkin, Pumpkin by Jean Titherington
large pumpkin vine poster tacked low on a wall
one small paper pumpkin for each child
blindfold
pumpkin seeds
small vegetable pumpkin

Group Size

The whole class may participate.

Activity Agenda

Gather children around you and without telling children what kind of seeds you have, pass out one pumpkin seed to each child. Let them guess what they think the seed will grow into. Record the guesses on the chalkboard or a chart. Collect the seeds and tell the children you will read them a story that will describe what this kind of seed will grow into.

Ask how many children have ever played Pin the Tail on the Donkey. Explain that this new game is called Put the Pumpkin on the Vine, and it is played in the same way. In this game, each child will have a turn to put a small pumpkin on the vine poster.

Ask for a volunteer. Place the blindfold over the child's eyes. Spin the child around three times and point the child in the direction of the vine poster. Give the child a small paper pumpkin with a piece of tape extending from the stem. When the child reaches the poster, have him or her attempt to put the pumpkin on the vine poster.

Explain that some children will end up putting their pumpkins on the vine and others will put their pumpkins in the pumpkin patch. Children should feel successful no matter where their pumpkin is placed.

Extended Activities

- Count and see how many children pasted their pumpkins on the line. How many children missed the vine and put their pumpkins in the pumpkin patch? Were their more pumpkins on the vine or in the pumpkin patch?
- Discuss the sequence of the story. In order for a pumpkin to grow, what happens first? What happens next? How long does it take from the time the pumpkin seeds are planted until the pumpkin is full grown? Ask other questions to show a growing time line.
- Set up a library center with pumpkin props and books such as: *Apples and Pumpkins* by Anne Rockwell.

Garden and Seed Puzzles

Children will use fine motor and problem-solving skills by doing puzzles while recognizing things that grow in the garden.

Materials

variety of wood and paperboard puzzles–ranging from easy to more difficult–featuring vegetables, fruits and flowers

large copy of a seed puzzle using the pattern on the right. Enlarge this seed art onto a piece of heavy tagboard. Cut it into as many puzzle pieces as there are children in your class.

Group Size

The whole group of students can participate in the beginning puzzle activity. Small groups of four to six students can work at the puzzle center.

Activity Agenda

Gather the children in front of you and give each child a piece of the seed puzzle without telling them what the puzzle will form. Have students guess what all the puzzle pieces could make. Have children work together on the floor and complete the puzzle to form the seed.

Assign children, a few at a time, to work on the individual fruit, flower and vegetable puzzles. As children complete the easier puzzles, advance them to more difficult ones.

Extended Activities

- Read *A Seed Is a Promise* by Claire Merrill.
- Have a puzzle contest! Who can finish a puzzle first?
- Ask parents to donate a cake or bread made with fruits or vegetables. See if children can guess what special garden ingredient was used. Suggestions may include: carrot cake, banana bread, zucchini bread, pumpkin bread.

More Spatial Activities

Up Above or Down Below

Make a set of vegetables using colored felt. Provide a background of a large piece of brown felt or construction paper and make a horizontal line to divide it into two sections: *above the ground* and *below the ground*. Have children practice placing the vegetables where they grow: *up above* or *down below*. Have children repeat the rhyme:

> "Put them in the place they grow,
> Up above or down below."

Fruit and Vegetable Concentration

Make a set of fruit and/or vegetable playing cards with matching pairs of each. Play Concentration by matching pairs. Fruit and/or vegetable stickers and index cards make a quick game.

Playdough

Provide children with different colors of playdough, cookie cutters, small plastic fruits and vegetables. Have them make fruits and vegetables and/or baskets to keep the fruits and vegetables in.

Pattern Block Flowers

Provide children with pattern blocks to make flowers. Set a large piece of butcher paper under the area where the children will design flowers. Show children how to "save" their flower design by tracing around their pattern and then coloring it.

Make a Flower Garden

Purchase several inexpensive stems of plastic flowers. Cut individual flowers from each stem to provide more flowers. Set them in a basket. Provide a tray of soil and let small groups of children arrange the flowers to make a pretty garden. Encourage children to be creative in their designs.

Seed Planting

To help students understand how and why things grow; to encourage students to keep track of the changes happening to their seeds and to be responsible for caring for the seed plant.

Materials

individual student peat pots (many nurseries will donate these to schools if you ask them)
variety of fruit, flower and vegetable seeds that show growth quickly
pencils
rulers for measuring
seed logs or diaries for each student
Make individual seed logs for each student by taking four pieces of white paper and folding them in half. Staple at the top to make a booklet.

Activity Agenda

Gather the children around you in a circle. Review the different things that grow from seeds and what a seed needs in order to grow (sunshine, water, usually soil). Demonstrate how the students will plant a seed in a peat pot at the seed planting center.

Put student's name on his or her peat pot. Fill pot with soil. Take a pencil and poke a hole in the soil. Place a seed in the hole and cover the dirt again. Have the students make at least two holes and plant two seeds in the container.

Record what each child planted on a chart; then discuss what each child planted.

After students are finished planting, have them pick up a seed log, and write their names on the front of it and design the cover. Each day the students are to look at their seed, see what is happening and record it in the log. Measure the plant as it grows and record the growth information, also.

The Seeds Inside

Children will explore gardens and seeds by matching seeds to vegetables, fruits and flowers that grow in gardens.

Materials

variety of seeds in small, clear plastic bags (select seeds that are similar but can be differentiated by students if they look closely)

cards showing a fruit/flower/vegetable on one side and the seed it grew from pasted on the back

Group Size Small groups of five to six at one time

Activity Agenda

Gather the children in front of you. Explain that you will be matching seeds to pictures of fruits, flowers and vegetables.

Show a card with a picture or drawing of a watermelon (or slice) on one side and watermelon seeds pasted on the back. Show the various plastic bags filled with seeds that correspond to the different cards.

Working with children in small groups, let each child select a fruit/flower/vegetable card and study the picture on one side and the seed on the other. Ask students to find the bag of seeds that matches the seed on their card. Encourage children to study the size, color and shape of the seed before selecting the seed bag. Repeat the activity allowing children to match several seed cards to seed bags.

Extended Activities

- Let children examine the seeds with magnifying glasses.
- Ask children to describe the seeds. Encourage words emphasizing size, shape, color, texture.
- Have children dig seeds out of vegetables and fruit with plastic utensils, and let them create their own picture/seed cards.

More Natural Science Activities

Trees in Our Neighborhood

Take a walk around the school and notice the different trees. Which ones can you name? Point out common types such as oak, elm, maple, birch, pine.

Seed Sorting

Gather seeds outside, from seed packets and from seed stores and put them into a center with sorting containers. See how many different ways the students can sort them.

Nature Walk

Take a nature walk. Provide students with paper or pencils so they can record what they see.

The Tiny Seed

Read the book *The Tiny Seed* by Eric Carle and talk about the ways seeds travel, are scattered or broadcast. (*Broadcast* means "to scatter over a broad area.") See how many different kinds of seeds the children notice.

Made from Maple

Visit the library and select some children's books that clearly describe and show the sugaring process of how sap is made into maple syrup, *Sugaring Time* by Kathryn Lasky is excellent. Bring in maple syrup and have children sample. Set up a display of books and encourage children to learn about this fascinating "tree-tapping" process in New England. View videotapes to show the sugaring.

The Life Cycle of Trees

Depending on the resources in your immediate school environment, arrange a visit to a tree nursery or local children's museum to learn about the various stages in a tree's life. Have children see a "slice" of a tree with rings that give clues as to its age.

What Goes in Vegetable Soup?

Students will listen to a story and practice personal and social development by participating in a group activity; students will make logical choices based on the story.

Materials

Growing Vegetable Soup by Lois Ehlert
large pot or roasting pan
cut out photographs of foods that are frequently used in making vegetable soup as well as photographs of foods that are not included in vegetable soup such as: cottage cheese, coffee, cereal, cookies, bananas, ice cream, etc.

Group Size

The whole class may participate.

Activity Agenda

Gather the students in front of you and read *Growing Vegetable Soup*. After discussing the story, show the children the large soup pot or roasting pan and the photographs of the different foods. Name each food along with the children. Pass out one food photograph per child.

Tell the children that each one will name the food in their picture and put it "in the pot" if it is something that is used in cooking vegetable soup.

Take turns and go around in a circle giving each student an opportunity to name his or her food and explain why it is or isn't used in cooking vegetable soup. Sample answers might be: corn is used in vegetable soup because it grows from the ground and is a vegetable; chicken can be used along with vegetables in vegetable soup because my grandma makes it that way and it's good; or ice cream can't be used in vegetable soup because it would melt and the soup would taste funny. Accept all logical answers and have children enjoy the humor that may come from this activity!

Extended Activities

- Look and see what pictures are in the pot. List the ingredients on chart paper.
- Assign each child to bring in a specific ingredient and cook vegetable soup, or bring in cans of vegetable soup, heat and serve.
- Read *Stone Soup* by Marcia Brown and discuss the virtues of cooperation.

Fruit and Vegetable Alphabet

To make students aware of different kinds of fruits and vegetables that grow while reinforcing counting, sorting and matching skills; to have children taste new foods.

Materials

Eating the Alphabet by Lois Ehlert
samples of common fruits and/or vegetables along with some new unfamiliar ones
large graph

Group Size

The whole class may participate.

Activity Agenda

With the group seated in front of you, ask children to name some fruits. Then ask children to name some vegetables. Record their responses on a chart divided into fruit/vegetable columns. Show the children the book *Eating the Alphabet* and explain that the book has many different fruits and vegetables, and the author has used the entire alphabet.

Read the story and explain unfamiliar items. After reading, allow children to discuss their favorite fruits.

Prepare a large graph and let children suggest which fruits should be listed based on previous responses. Invite each student to come up and color in a square to show his or her favorite fruit. What is the most popular fruit? What is a favorite summer fruit? (Repeat this activity on a different day using vegetables.)

Directions

- Have children bring in fruits and prepare fruit salad. Add one or two unusual fruits from the market and let the children find it in the salad.
- Make a list of fruits that are found only in the summer.
- Visit a local orchard, fruit farm or commercial grower and see what fruits are in season.
- Buy some berries in season and make fruit sherbet.
- Set up a fruit/vegetable stand and let children play/act shopping for produce.

More Nutrition Activities*

Taste and Describe

Bring in a fruit or vegetable that you think children have not tasted before. Cut it into small pieces and let children sample the food. What does it taste like? How would you describe it? Use descriptive words such as *crunchy, smooth, sweet, sour, bitter, salty.* (Jicama is an excellent vegetable for this activity. It looks like a cross between a potato and a turnip and is eaten raw. When peeled and cut, it tastes similar to an apple. It is crunchy and good in salads. Ask the produce manager at your local market.)

Delicious Dirt

Have some fun with this nonnutritional activity! Let children make chocolate pudding–either the cooked or instant type. Divide into small cups. Crumble chocolate cookies on top. Put gummy worms in the cups. Enjoy eating delicious dirt!

Have a Hawaiian Salad

Bring in books about Hawaii and learn about the different crops grown there. Prepare a fruit salad using pineapple, coconut, mango and papaya. Play Hawaiian music during this activity.

New Fruits, New Tastes

Make an illustrated chart showing new fruits and vegetables as children are introduced to them. Try rhubarb, kiwi, jicama, plucots.

*Be aware of any food allergies your students may have and closely supervise the use of small items.

Student Reproducibles

To the Teacher

Page 56 can be used to introduce this unit. Students may write or dictate two to three facts about gardens and seeds. This activity may be done in small groups. Save completed pages to compare with the fact sheets students will complete at the end of the unit.

Use page 57 to determine each student's individual goals. Ask children to think about what they would like to learn about gardens and seeds before distributing the page. Students may write or dictate three to four things they would like to learn about seeds and gardens. Save completed pages to compare with final page of learned facts.

Page 58 can be used at the end of the unit. Ask children to read page 57 "This is what I want to learn about gardens and seeds," and see if they learned what they set out to learn.

The form on page 59 can be used to evaluate the students' knowledge of gardens and seeds.

Use the open-ended grid on page 60 to make graphs with your class such as *Our Favorite Fruits, Our Favorite Vegetables* or *Our Favorite Flowers.* You may fill in the title and send the page home with children to complete with their family members.

Page 61 can be used to reinforce counting and adding. Make similar math pages with more difficult counting and adding activities.

Use page 62 to write new vocabulary or spelling words; to use for math facts or for children to illustrate fruits, vegetables and flowers they are learning about.

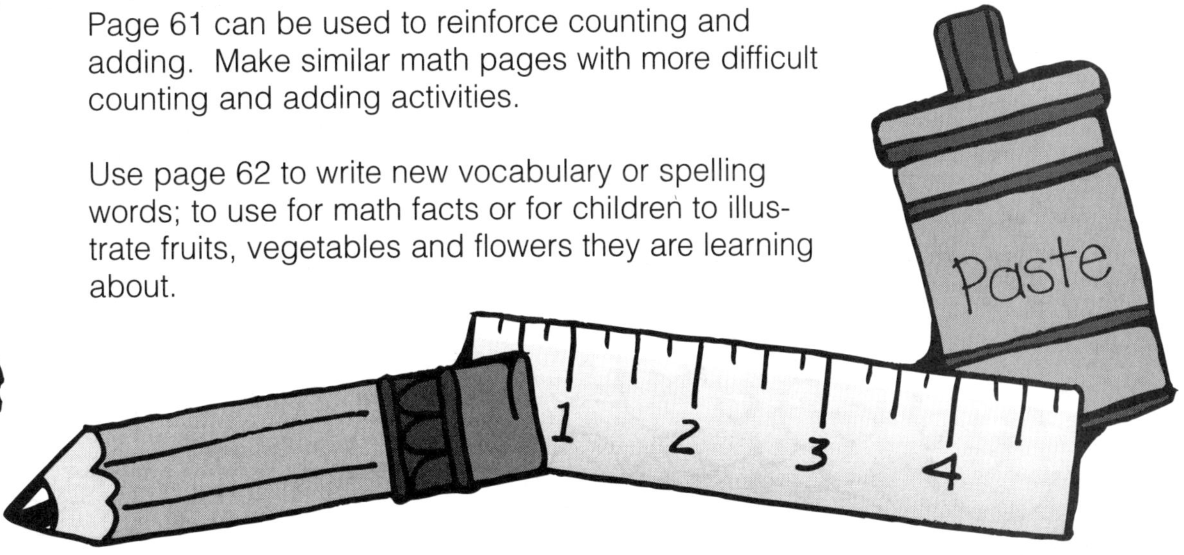

Name ______________________________

This is what I already know about gardens and seeds.

Name ____________________

This is what I want to learn about gardens and seeds.

Name ______________________________

This is what I have learned about gardens and seeds.

____

____

____

____

____

____

____

____

____

____

____

Name ______________________________

I Know! I Know! How Gardens Grow!

_____ I know what seeds need to grow.

_____ I can plant a seed.

_____ I know what will happen when a seed is planted.

_____ I know what seeds are.

_____ I can match some seeds to plants.

_____ I know several types of gardens.

_____ I know how to take care of a garden.

_____ I know some vegetables in a vegetable garden.

_____ I know some flowers in a flower garden.

_____ I know some animals that like gardens.

Name ______________________________

Our Favorite ______________________

Name ______________________________

Garden Math

1. 

2. 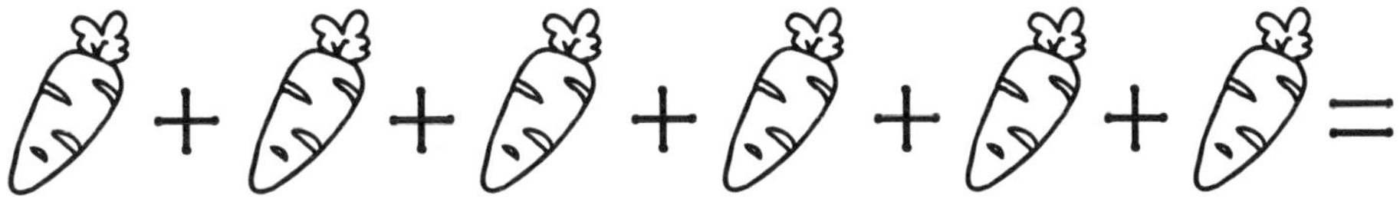

3. 

4. 

5. 

6. 

7. 

8. 

Name ______________________________

Suggested Literature for Children

Birney, Bett. *Oh, Brother! Someone Won't Share.* Racine WI: Western Publishing, 1993.
Rabbit won't share, but the animal friends come to help protect his garden anyway. Rabbit decides that sharing is a good thing to do.

Brown, Marcia. *Stone Soup.* New York: Simon & Schuster Children's, 1947.
Three resourceful soldiers show some miserly townspeople how to turn a stone into a meal fit for a king!

Brown, Margaret Wise. *The Runaway Bunny.* New York: HarperCollins Children's Books, 1977. Other editions are available.
A little bunny's mother reassures him that no matter where he goes, she will always find him and love him.

Carle, Eric. *The Tiny Seed.* New York: Scholastic Inc., 1987.
Colorful illustrations and dramatic language trace the cycle of seeds.

Cole, Henry. *Jack's Garden.* New York: Greenwillow Books, 1995.
Patterned after "The House That Jack Built," this clever garden version shows how Jack grows a beautiful flower garden complete with birds, bugs and butterflies.

Cristini, Eric. *In My Garden.* Austria: Veriag Neugebaucer Press, 1981.
A picture book shows large and small creatures that make their home in the garden.

Davenport, Zoe. *Words for Everyday: Garden.* New York: Houghton Mifflin, 1995.
Simple illustrations of plants, bugs, birds and tools contain one-word labels.

Ehlert, Lois. *Eating the Alphabet.* San Diego: Harcourt Brace Jovanovich, 1988.
This is a great alphabet book. It lists vegetables and fruits that begin with each letter of the alphabet. It also has a wonderful glossary in the back that tells about each of the fruits and vegetables.

Ehlert, Lois. *Growing Vegetables.* New York: Scholastic Inc., 1987.
This delightful book shows the steps to planting, gardening, picking vegetables and making vegetable soup.

Ehlert, Lois. *Planting a Rainbow.* New York: Scholastic Inc., 1988.
The story shows how planting a variety of flowers can look like a rainbow of color.

Faulkner, Matt. *Jack and the Beanstalk.* New York: Scholastic Inc., 1965.
Jack trades the family cow for some magic beans. The beans grow into a beanstalk and Jack goes to visit a giant.

Heller, Ruth. *The Reason for a Flower.* New York: Scholastic Inc., 1983.
Colorful illustrations and language show the life cycle of plants and ways other living creatures help flowers spread their seeds.

Johnson, Paul Brett. *Farmer's Market.* New York: Orchard Books, 1997.
A beautifully illustrated modern-day story of two young girls who spend Saturday's together at a bustling farmer's market where Laura's family sells their vegetables and Betsy's mothers sells flowers. Large fold-out pages bring children right into the scene of the farmers' market.

Jordan, Helene J. *How a Seed Grows.* New York: HarperCollins, 1992.
A Let's Read and Find Out Science book show pictures about seeds and plants.

Krauss, Ruth. *The Carrot Seed.* New York: Scholastic Inc., 1945.
A small boy plants a carrot seed. Patience and hard work result in a big surprise.

Kroll, Steven. *The Biggest Pumpkin Ever.* Holiday House, 1984.
The pumpkin-growing adventures of two different mice.

Lasky, Kathryn. *Sugaring Time.* New York: Simon & Schuster Children's, 1984.
Beautiful black and white photographs accompany a lyrical text about a Vermont family making syrup. Newbery Honor Book.

Lobel, Anita. *The Rose in My Garden.* New York: Scholastic Inc., 1984.
Picture by picture this book builds a flower garden.

Marzollo, Jean. *I'm a Seed.* Cartwheel, 1995.
Two seeds grow side-by-side a marigold and a pumpkin.

McCloskey, Robert. *Blueberries for Sal.* New York: Scholastic Inc., 1976.
Sal and her mother pick blueberries and meet a bear.

McMillan, Bruce. *Eating Fractions.* New York: Scholastic Inc., 1991.
Colorful photos illustrate two children's food cut into halves, quarters and thirds and how those parts make a whole.

Medearis, Angela. *Picking Peas for a Penny.* New York: Scholastic Inc., 1990.
This book tells about how much people got paid for picking peas on a plantation.

Merrill, Claire. *A Seed Is a Promise.* New York: Scholastic Inc., 1990.
Wonderful illustrations show how it is the seed's promise to grow into what it is supposed to grow into.

Moore, Elaine. *Grandma's Smile.* Lothrop, Lee and Shepard, 1995.
Kim's grandmother always has a smile when she comes to visit, and it is the inspiration for Kim's jack-o'-lantern face.

Parnell, D. *The Little Red Hen.* New York: Ladybird Books, 1984.
The little red hen finds some grain and no one will help her plant it, cut it, or take it to the mill. So when she makes it into bread, she eats it all by herself.

Poncavage, J. *Grow a Totally Weird Garden.* Canada: Becker and Mayer, 1996.
A guide to a complete garden including plants, bugs, seeds, leaves and harvest.

Potter, Beatrix. *The Tale of Peter Rabbit.* New York: Scholastic Inc., 1986.
Peter Rabbit disobeys his mother and runs into trouble.

Raffi. *Everything Grows.* New York: Crown Publishers, 1989.
Excellent pictures and a song tell about many things that grow.

Rockwell, Anne. *Apples and Pumpkins.* New York: Scholastic Inc., 1989.
A family takes a trip to an orchard to pick apples and pumpkins and then take them home to use them.

Rowe, Julian, and Perham, Molly. *Watch It Grow.* Chicago: Children's Press, 1994.
A book of photographs showing how to plant and watch things grow.

Saskie, Judy. *The Seed Song.* California: Cypress Press, 1994.
A song that tells about how seeds grow.

Simont, Marc. *A Tree Is Nice.* New York: Harper & Row, 1956.
A classic story about the value of trees.

Steele, Mary. *Anna's Garden Songs.* New York: Scholastic Inc., 1989.
From beets to nasturtiums, Anna tours the family garden with songs about growing, picking and eating wonderful plants.

Titherington, Jean. *Pumpkin, Pumpkin.* Greenwillow, 1986.
Jamie plants a pumpkin seed and watches it grow. After carving it, he saves some seeds to plant in the spring. Beautiful illustrations.

Wexler, Jerome. *Flowers Fruits & Seeds.* New York: Simon & Schuster, 1987.
This book contains pictures of flowers and fruits and their seeds.

Wilkes, Angela. *My First Garden Book.* New York: Alfred A. Knopf, 1992.
A life-sized guide to growing things.

Zolotow, Charlotte. *Mr. Rabbit and the Lovely Present.* HarperCollins Children's Books, 1977.
Mr. Rabbit helps a little girl find a present for her mother.